LIFE
SCIENCE
STORIES

Plant Classification

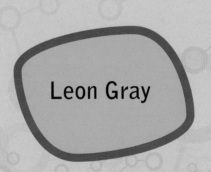

Leon Gray

Gareth Stevens
Publishing

Please visit our website, www.garethstevens.com. For a free color catalog of all our high-quality books, call toll free 1-800-542-2595 or fax 1-877-542-2596.

Library of Congress Cataloging-in-Publication Data

Gray, Leon, 1974-
 Plant classification / Leon Gray.
     p. cm. — (Life science stories)
 Includes index.
ISBN 978-1-4339-8720-5 (pbk.)
ISBN 978-1-4339-8721-2 (6-pack)
ISBN 978-1-4339-8719-9 (library binding)
1. Plants—Juvenile literature. 2. Plants—Classification—Juvenile literature. I. Title.
 QK49.G647 2013
 581—dc23

                              2012023552

Published in 2013 by
Gareth Stevens Publishing
111 East 14th Street, Suite 349
New York, NY 10003

© 2013 Gareth Stevens Publishing

Produced for Gareth Stevens by Calcium Creative Ltd
Designed by Paul Myerscough and Geoff Ward
Edited by Sarah Eason and Harriet McGregor

Picture credits: Cover: Shutterstock: Cozyta (r), Miroslav Hlavko (l), Jurasy (bg). Inside: Shutterstock: 5r, A9 Photo 26, Alexnika 10, Scisetti Alfio 21, Artincamera 14, Ainars Aunins 18, Alena Brozova 20, Oleksandr Bystrikov 22, Carlos Caetano 7cl, Ewan Chesser 29, Ckchiu 11, Collpicto 15, Cozyta 27, Formiktopus 8, Daniel Gale 24, Roman Gorielov 16, Jubal Harshaw 28, Natalie Jean 12, Panos Karapanagiotis 7, Christopher Kolaczan 9, D. Kucharski & K. Kucharska 17, Marykit 23, MTR 6, Lee Prince 25, Matt Tilghman 4, Ulkastudio 9cl, Vilor 5l, VVO 19, Wallentine 13.

Printed in the United States of America

CPSIA compliance information: Batch #CW13GS: For further information contact Gareth Stevens, New York, New York at 1-800-542-2595.

Contents

Our Living World

Our world is packed full of many living things, called organisms. From bees to lizards and bats to human beings, living things are everywhere on Earth.

Living Groups

Scientists group living things by the way they look and live. This is called classification, and it helps people make sense of nature. Scientists have aleady found around 10 million different living things, and they are finding more every single day.

There are five main groups of living things, called **kingdoms**. They are animals, plants, fungi (such as mushrooms), **bacteria**, and **protists**.

Many lush, green plants grow in wet places such as this.

HOW MANY PLANTS?

No one knows exactly how many different plants there are in the plant kingdom. Plant scientists (called botanists) guess that there are between 320,000 and 425,000 plants. Every year, the total number of plants grows. Botanists find new plants in places such as tropical rain forests and cold mountaintops.

Many different plants live in cold and wet places such as alpine meadows.

The Story of Classification

The story of classification began around 2,300 years ago. A Greek scientist named Aristotle (384–322 BC) grouped living things as animals or plants. Aristotle's classification system lasted until the eighteenth century.

Even Smaller Groups

Carolus Linnaeus (1707–1778) grouped living things into smaller groups of creatures that are alike. He called these groups "genera" and the unique living things within each group "species." Linnaeus used a language called Latin for group and species names, because most scientists spoke Latin.

Studying plants in the laboratory helps scientists to classify species.

This is a statue of Aristotle. He was the first person to classify living things hundreds of years ago.

True Story

THE FIRST BOTANIST

Aristotle is most famous for his writings about animals. One of his students, named Theophrastus, was the first person to classify plants.

Food Factories

Plants make their own food. To do this, they use energy from sunlight. This is called photosynthesis.

Food from Light

Plants also need carbon dioxide and water to make food. Plants take in carbon dioxide from the air through their leaves. They suck up water from the ground through their roots. **Chlorophyll** is a green substance found in plant leaves, and it traps energy from sunlight. During photosynthesis, the trapped energy turns carbon dioxide and water into sugar. This sugar is the food all plants need to grow.

The sunflower is named after its big, yellow flowers.

PLANTS FOR LIFE

Without plants, no life on Earth would exist. Animals cannot make their own food. They need plants for food. Even animals that eat only meat need plant-eating animals for food. In photosynthesis, plants release oxygen into the air, which is breathed by all living things.

The American bison grazes on grasses on the prairies of the American Midwest.

The Plant World

Earth's plant kingdom is split up into small groups. The plants in each group have similar features, such as flowers or cones, or grow in a similar way.

How Do We Group Plants?

Vascular plants have stems and leaves through which water can pass. Vascular plants can be divided into even smaller groups. One group of plants uses **spores** to **reproduce.** The other group uses seeds instead.

Water cannot flow through some plants. Instead, these plants soak up water from their surroundings. These plants are known as nonvascular plants.

Orchids are flowering plants with pretty, colorful flowers.

GIANT SEQUOIA

The giant sequoia is one of the largest of all plants. This tree can grow more than 300 feet (90 m) tall. It can live for more than 3,000 years. Sequoias are conifers, which means they reproduce by using seeds in cones.

The giant sequoia grows on the slopes of the Sierra Nevada mountains in California.

Amazing Algae

Algae are simple plantlike organisms that live in water and damp places. Water animals eat algae. These organisms are so unusual that scientists have put algae in a group of their own.

Shapes and Sizes

Algae come in many shapes and sizes. Phytoplankton are tiny and live in the oceans. They are so small that 1,000 could fit on the period at the end of this sentence. Larger algae, called seaweed, grow in huge underwater forests. These can stretch from the seabed right up to the ocean's surface.

Seaweed fronds may look like plant leaves, but seaweed and plants are completely different organisms.

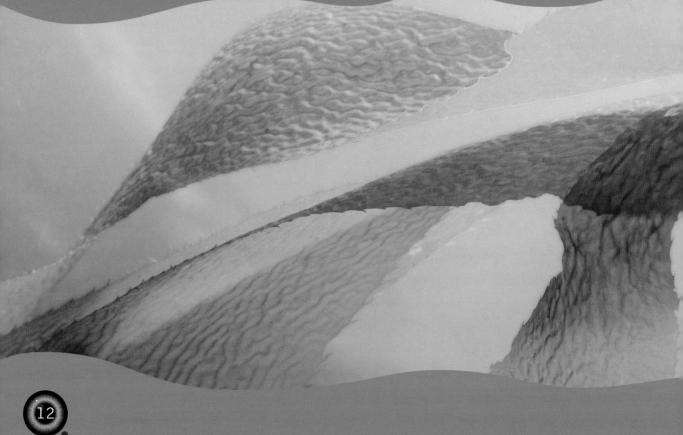

PLANTS OR NOT?

Green algae often look like plants, but other algae are very different. The algae in the hot springs of Yellowstone National Park live in temperatures of more than 160°F (70°C). These superhot temperatures would kill any plant on Earth.

The bright color of the water in this hot spring comes from its algae and bacteria.

Mosses and Liverworts

Mosses and liverworts are nonvascular plants. They do not make seeds or flowers. Instead, they reproduce from tiny spores. These spores grow inside capsules and blow away in the wind.

Different Kinds of Survival

Mosses often grow as "carpets" of plants on rocks. They are covered with a waxy coating, which stops them from losing water. Liverworts live in damp soil, on logs and rocks, and in water. Tiny strands, called **rhizoids**, fix the plants to the ground, and they also suck up water.

Moss grows on almost everything on a forest floor. Unlike most other plants, mosses have no leaves, roots, or stems.

The flat parts of the liverwort look like the shape of a human liver! This is why botanists called it liverwort.

WATER LOVER

The liverwort can survive only in very wet places. In the wild, it grows on damp rocks and soil near bogs, rivers, and streams. The liverwort is also a weed. It grows very well in the damp ground near ponds.

Other Vascular Plants

Club mosses, ferns, and horsetails are vascular plants. They do not make seeds, but use other ways to make new plants.

Thousands of Plants

There are about 10,000 different ferns. Ferns make light spores, which blow away in the wind. Spores grow into new plants where they land. Club mosses are like mosses, but have leaves, roots, and stems. There are about 1,000 different club mosses. There are only about 30 different horsetails. Some are tiny and just a few inches high. Others, such as the giant horsetail, can grow up to 16 feet (5 m) tall.

The leaves of a fern are called fronds. Young ferns have tightly curled fronds. The fronds slowly uncurl as the plants grow.

Horsetails reproduce from conelike shapes that grow at the end of their stems.

True Story

LIVING FOSSILS

Horsetails and ferns are sometimes called "living fossils." This means that they have changed very little in the last 100 million years since they first appeared on Earth. Millions of years ago, there were huge forests of ferns and horsetails. They were eaten by enormous plant-eating dinosaurs.

Seed Plants

"Seed plants" make up about 92 percent of all plants on Earth. They include plants with cones and all flowering plants. Seed plants reproduce using seeds.

Spore Versus Seed

Spores are so small that they can only be seen under a microscope. The microscope magnifies the spore—making it look bigger. Spores only grow into new plants if they land on wet ground.

Seeds are tougher than spores. A seed has a covering called a **testa** which protects the delicate parts inside. It also contains food for the plant.

When a bird eats the fruits of a plant, it spreads the seeds onto land in its waste.

SPREADING SEEDS

Seeds help plants grow in new places. Plants use different ways to spread their seeds. Small, light seeds may be carried by the wind or they may wash away in rainwater. Other plants use animals to spread their seeds in their waste.

Some large seeds, such as coconuts, can float. They drift on ocean currents and grow into plants when they land on new islands.

Plants with Cones

Conifers and similar plants reproduce using seeds. The seeds are on the cones of the plant and may be male or female. Male cones contain pollen. Female cones contain eggs. A seed forms when pollen joins with an egg, and then grows into a new plant.

The Family Conifer

Three groups of plants are related to conifers. These are cycads, ginkgos, and gnetophytes. Cycads are tropical plants that look like palm trees. Ginkgos are a type of plant that grows in China. Gnetophytes are woody plants that live in deserts and hot tropical places.

The large cone of this conifer holds the seeds that will grow into new plants.

ANCIENT PLANT

Ginkgos are also known as maidenhair trees. They are the only living species of a group that has been around for 270 million years. Ginkgos have fan-shaped leaves and seeds that smell gross! People grow ginkgos for food and also use them in medicine.

Scientists believe that medicine made from the leaves of ginkgos may help people to think and even remember things!

Plants with Flowers

Most plants are flowering plants. There are at least 230,000 species of flowering plant. The seeds of flowering plants are made inside an **ovary**. The ovary is hidden in flowers or fruits.

What's in a Flower?

The flower is the reproductive part of a plant. All flowers grow from a bud. Some plants have flowers with male parts, called **stamens**. The stamens produce pollen. Other plants have flowers with female parts. These parts include the egg inside the ovary, the **stigma**, and the **style**. Some plants can have both male and female flowers on the same plant.

The red flowers of the poppy contain both the male and female parts of the plant.

DESERT LIVING

Cacti produce flowers and seeds quickly during a desert's short rainy season.

Cacti are flowering plants that live in deserts and other dry places. They are suited to life where there is little water. Cacti have spines instead of leaves. The stems get bigger to store water, and cacti use this water when there is little rainfall.

Pollination

When pollen travels to the female part of a flower it is called **pollination**. Plants use two types of pollination.

The Cycle of Life

Self-pollination is when pollen from a male flower passes to the female flower of the same plant. When pollen from one plant passes to the female flower of another it is called cross-pollination. Pollen lands on the stigma and a tube grows into the style. The pollen moves down the tube. It joins with the egg in the ovary. This is called **fertilization**.

When bees carry pollen from one plant to another, cross-pollination occurs.

True Story

THE STRANGLER

A tree called the strangler fig kills other plants to survive. It grips onto another plant, called a "host," then wraps its roots around the plant to strangle it. The strangler fig takes the dead plant's place in the forest.

The woody roots of the strangler fig grow around its host. They cut off its food supply.

Meat-eating Plants

Some plants eat animals such as frogs and insects! Most live in bogs and swamps. Meat-eating plants get most of their food through photosynthesis. They just top off missing **nutrients** by eating small animals.

Killer Plants

There are two main groups of meat-eating plants. Some plants, such as the Venus flytrap, have leaves that snap shut to catch prey. Other plants trap **prey** by using sticky liquids, suction, or water pools inside them. Once the prey is trapped, the plant's chemicals break down the animal's body.

The leaves of the Venus flytrap snap shut when insects touch the trigger hairs between them.

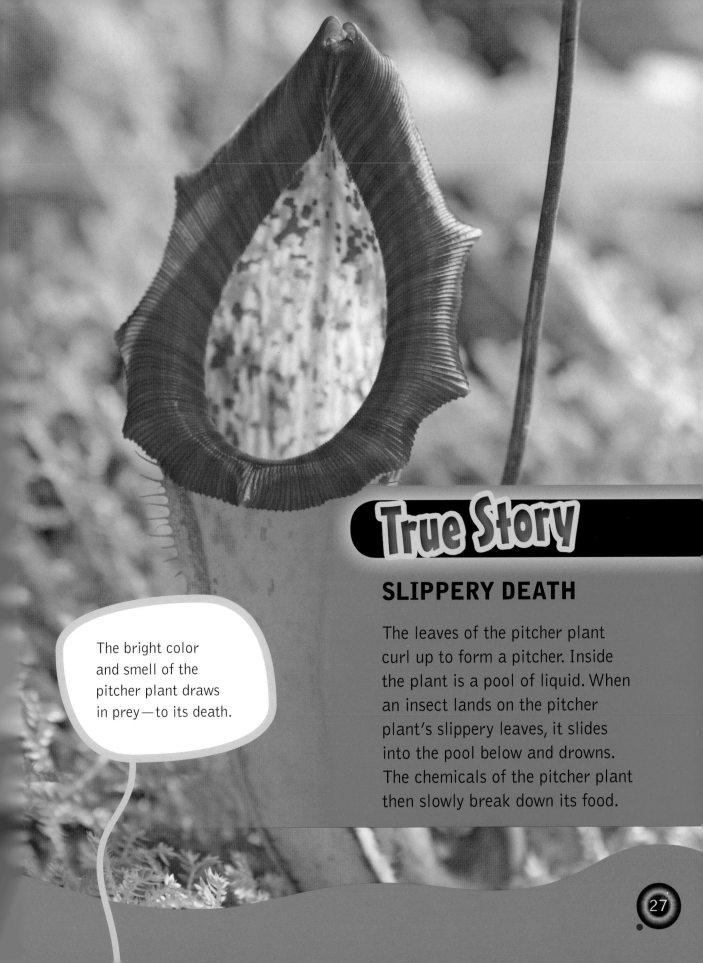

The bright color and smell of the pitcher plant draws in prey—to its death.

SLIPPERY DEATH

The leaves of the pitcher plant curl up to form a pitcher. Inside the plant is a pool of liquid. When an insect lands on the pitcher plant's slippery leaves, it slides into the pool below and drowns. The chemicals of the pitcher plant then slowly break down its food.

Classification Today

Botanists once used the way a plant looked to classify it. Things are much easier today. Botanists now look at the **DNA** of a plant. DNA is a code that controls everything about the plant, from its shape to its color.

Clues Through DNA

Today, botanists compare the DNA of plants to figure out how plants are related and how they developed. They have shown that some plants are more or less closely related than people first thought. Botanists change plant classification according to these findings. This helps to record new species and protect plants from dying out.

Plant **cells** contain DNA. Botanists study them to learn about the relationships between plants.

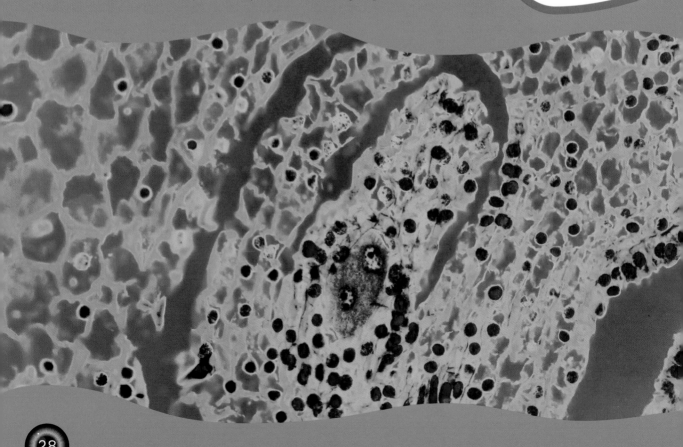

RAPESEED AS FUEL

Some plants are used as a source of "**biofuel**" for motor vehicles. Canola oil comes from the seeds of the rapeseed plant. The oil is usually mixed with normal diesel fuel. It has been used to power buses, trains, and even jet aircraft.

The seeds of these rapeseed plants will be turned into biofuel to power cars and other vehicles.

Glossary

bacteria: single-celled organisms that are neither plants nor animals

biofuel: any fuel that is produced from living things, such as plants

cell: the smallest part of any living organism

chlorophyll: the green substance that helps plants to make food

DNA: deoxyribonucleic acid—the molecule that decides how living things will look

fertilization: when a male cell joins with a female egg to make a new plant

kingdom: large groups of organisms that are related

nutrient: part of food that organisms need to live and grow

ovary: part of the female plant in which eggs are produced

pollination: the transfer of male pollen to the female stigma of a flowering plant

prey: animals that are eaten by other animals

protist: microscopic organism that is neither plant nor animal

reproduce: to make a new plant or animal

rhizoid: threadlike hair on the bodies of plants such as mosses and liverworts

spore: the light part of a seedless plant that can be carried by wind to another area of land where it grows into a new plant

stamen: the male reproductive part of a flower

stigma: the part of the female reproductive system of some plants that collects pollen

style: the stalk that supports the stigma

testa: the protective outer layer of a seed

For More Information

Books

Ballard, Carol. *Plant Variation and Classification.* New York, NY: Rosen Central, 2010.

Burnie, David. *Eyewitness Plant.* New York, NY: DK Publishing, 2011.

Spilsbury, Louise, and Richard Spilsbury. *Plant Classification.* Chicago, IL: Heinemann Library, 2008.

Websites

Search this website for lots of information about plants and their structure, plant classification, photosynthesis, and much more.
www.mcwdn.org/Plants/PlantsMain.html

Find out more about plants by searching for them on the United States Department of Agriculture website.
plants.usda.gov/java

Play games, watch video clips, and view animated diagrams to learn about the structure of plants, photosynthesis, pollination, seed dispersal, and plant adaptations.
www.mbgnet.net/bioplants/main.html

Index